Cool Custom
Big Rig Trucks

A
Coloring Book
of
Trucking Delights

by
Nola R. Hintzel

Other Coloring Books by Nola Hintzel

Planet Geometry: Symmetrical Patterns of Proportion and Balance

Robot Fashion Show: A Geometrical Coloring Book Experience

Old Southwest Missouri: An Adult Grayscale Coloring Book

Nostalgic Northwest Arkansas Book 1

Nostalgic Northwest Arkansas Book 2

Agate Dewdrops: A Coloring Book for Everyone

Like Clockwork: A Coloring Book About Clocks

Historic Hot Springs Coloring Book 1: America's Spa City in Grayscale

Historic Hot Springs Coloring Book 2: America's Spa City in Grayscale

Cool Custom Big Rig Trucks: A Coloring Book of Trucking Delights

ISBN: 9781091932739

© Nola R. Hintzel

© Nola R. Hintzel

© Nola R. Hintzel

© Nola R. Hintzel

© Nola R. Hintzel

© Nola R. Hintzel

© Nola R. Hintzel

© Nola R. Hintzel

© Nola R. Hintzel

© Nola R. Hintzel

© Nola R. Hintzel

© Nola R. Hintzel

© Nola R. Hintzel

© Nola R. Hintzel

© Nola R. Hintzel

© Nola R. Hintzel

© Nola R. Hintzel

© Nola R. Hintzel

© Nola R. Hintzel

© Nola R. Hintzel

© Nola R. Hintzel

© Nola R. Hintzel

© Nola R. Hintzel

© Nola R. Hintzel

GRAYSCALE COLOR TEST PAGE

TEST YOUR COLORS

Made in the USA
Columbia, SC
22 August 2024

40947224R00057